BRIGHT IDEAS ABOUT COMMUNICATION

HEDY LAMARR

BY AZURA PEZZILLO

Hedy Lamarr had big ideas.

She loved to figure out ways to improve how things worked.

She was born a long time ago in a country now called Austria on the continent of Europe.

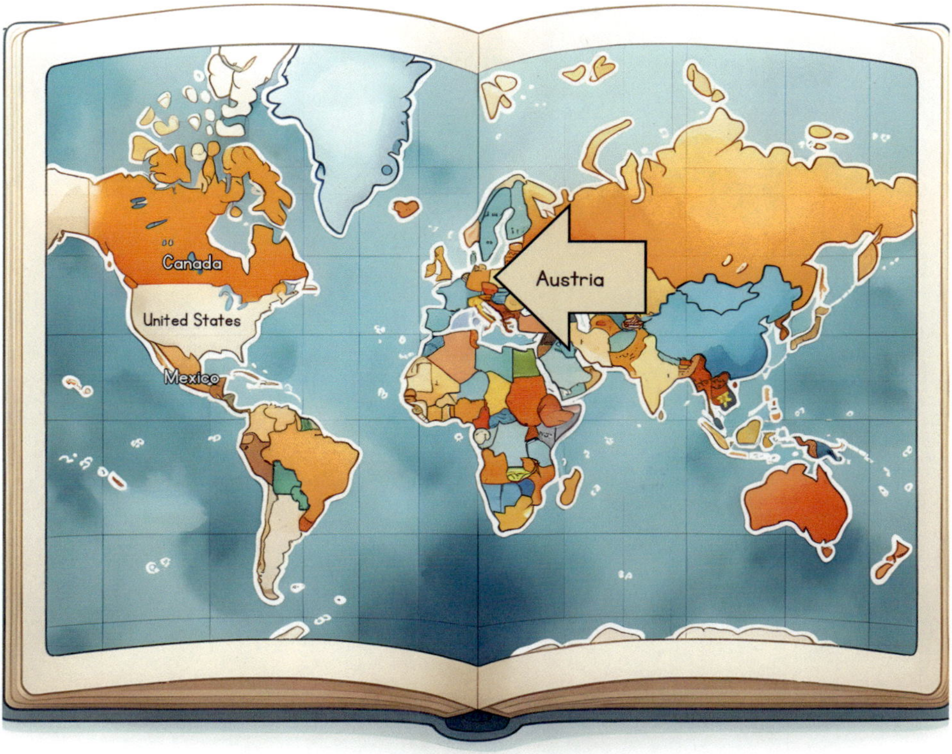
Canada
United States
Mexico
Austria

Even as a child, she imagined new ways to help people live better.

3+2=?
7-1=?
5×3=?

Hedy liked to learn, design, and be creative.

She became a famous movie actress in Europe.

DIRECT

She moved to the United States of America to work as an actress.

Sail to
New York

Hedy became a huge movie star in the United States of America. She made movies that people liked all over the world.

PREMIERE

Hedy found ways to create new inventions while working full-time as an actress.

Hedy worked with engineers to solve problems and design new products.

Aircraft Company

Hedy saw a problem with communication tools—and wondered how to make it better.

She tried and tried many things to come up with new inventions.

PROP
DEPT
SCENE
TAKE

She worked on her wireless communication invention with a music composer named George Antheil.

Together, Hedy and George came up with new ideas about how communication signals work.

After many tries, she and George invented "frequency hopping" to help prevent important communications from being blocked or stopped.

A
B

They donated the "frequency hopping" invention to the United States government for free because they wanted their ideas to help as many people as possible.

SIGNAL HOPPING
PATENT

Hedy's invention allowed other inventors to create mobile phone and wireless internet technologies like GPS, Bluetooth and WiFi.

Many people use Hedy's inventions to improve their lives every day.

Many engineers said “Wow” as they used George and Hedy’s invention to create new wireless tools.

Wow!
Wow!
1997
PIONEER AWARD

She helped make wireless communications easier, and more people could use them.

Hedy had bright ideas that changed the way we use wireless communication signals at home, at school, at work, and even in outer space!

Hedy Lamarr had ideas that changed the world.

Your ideas can change the world, too!

A B C

Want More?

Check out the books in the
Bright Ideas STEM Biography Series!

Reflection Questions for Parents and Educators

Here are fun questions you can ask children after reading the story together. There are no wrong answers — every idea matters!

These questions help children:

- Think like young inventors
- Explore creativity and teamwork
- Share big ideas about how the world works

Ask your child:

- What do you dream about?
- What do you like to create?
- What kinds of movies do you like?
- What job would you like to try one day?
- What do you like to make?
- Who do you like to work with?
- What do you like to do with friends?
- How do you help others?
- What would you like to invent?
- How could your invention help the world?

Glossary

Actress: A woman who performs in movies, plays, or television shows.

Austria [Awss-tree-ah]: A country in Europe where Hedy Lamarr was born.

Communication: Sharing information, ideas, or messages between people or devices.

Composer: A person who writes music.

Donated: Gave something as a gift to help others, without asking for payment.

Engineers [in-juh-neers]: People who design, build, and maintain machines, structures, and other things.

Frequency hopping [free-quin-see hop-ping]: A way to send messages by quickly switching signal paths, making them harder to block or interrupt.

George Antheil [An-thee-il]: A music composer who worked with Hedy Lamarr to invent frequency hopping.

Hedy Lamarr [Head-ee La-Mar]: An Austrian-American movie star and inventor with bright ideas about communication.

Inventor [in-vent-tor]: A person who creates something new to solve a problem.

Inventions [in-vent-shuns]: New devices, methods, or processes that someone has thought of and created.

Pioneer Award: An award given to people who create new and important ideas, often in science or technology.

Problem-solving: Finding ways to fix challenges or make things work better.

Signals: Waves or messages sent to convey information, such as sounds, lights, or electronic signals.

Technology: Tools, machines, or systems that people create to solve problems or make life easier.

Wireless: Working without wires, using signals sent through the air.

Wireless communication [Wire-less kuh-myoo-nuh-kay-shun]: Sending messages through the air without wires, like when you use a cell phone or a wireless internet connection.

Learning Connections for Parents and Educators

This book supports early science, technology, and engineering learning through storytelling, creativity, and real-world problem-solving.

It aligns with the **Next Generation Science Standards (NGSS)**, a widely used set of learning guidelines that help children build scientific thinking, collaboration, and design skills from an early age.

Classroom Discussion & Activities

Ready-to-use **class discussion questions and teacher notes** are available as a free downloadable resource for educators.

Go to **PreschoolPracticePress.com/classguide** to access the Bright Ideas Classroom Discussion Guide for this book.

These materials are designed to support group discussion, early engineering thinking, and student reflection in PreK–Grade 2 classrooms.

This book references the Next Generation Science Standards (NGSS) for educational alignment purposes only. This publication is not endorsed by, sponsored by, or affiliated with NGSS or its developers.

How This Book Supports Early STEM Learning

Exploring Problems & Solutions
Children are introduced to the idea that people notice problems and use creative thinking to improve how things work.

Engineering & Design Thinking
The story highlights how inventors and engineers design tools and technologies to help people communicate more effectively.

Signals & Communication
Young readers learn that messages can travel using signals, helping people share information without wires.

Cause, Effect & Systems
The book shows how changing one part of a system can improve communication for many people.

NGSS Connections (Grades K–2)

- K–2-ETS1-1: Asking questions and defining problems that can be solved through design
- K–2-ETS1-2: Developing ideas and solutions to improve how things work
- 1-PS4-4: Using signals to communicate over a distance

About the Author

AZURA PEZZILLO

Azura Pezzillo is a mom, wife, and chemical engineer who saw a gap in the preschool books she and her toddler read. She wanted more families to experience the fun part of Science, Technology, Engineering, and Math (STEM). Yes, that's right! There's a fun part to STEM learning!

Azura is passionate about helping curious children learn about our world when their brains are growing fastest - from age 0 to 6. With her new activity and biography book series, she's bringing the fun back to preschool learning. She strives to produce high-quality and well-researched children's books that give curious kids fun books and quality educational content - at the same time.

Azura lives with her family in New Jersey, close to the great museums, theaters, and libraries that inspired her to write stories and learn science as a child.

Follow on Amazon: https://www.amazon.com/author/azura

Published by Inman Oak Media on behalf of Preschool Practice Press
Email: hello@PreschoolPracticePress.com
Website: www.PreschoolPracticePress.com/BrightIdeas

First Edition: December 2025
Book design by Azura Pezzillo
Images used under license from Canva and OpenAI DALL-E
The illustrations in this book were rendered with generative artificial intelligence tools.

Made in the USA
Monee, IL
05 November 2026

67382333R00036